IMAGES
of America

VERMONT'S
MARBLE INDUSTRY

This is an early 1920s view of the Hollister quarry in Pittsford. Quarry workers descended many flights of stairs down hundreds of feet into the bottom of the quarries each day. As the stone was quarried, marble piers were left in place to support the weight of the walls. (Courtesy Proctor Historical Society.)

ON THE COVER: This early 1900s photograph shows the Vermont Marble Company at its peak production, with acres of marble blocks covering its grounds. As economic times and architectural styles changed, the fortunes of the company declined. By 1976, the Vermont Marble Company ceased to exit. Its operations were purchased by a multinational corporation that transformed the business from quarrying and finishing building stone to producing calcium carbonate in powder and slurry form for industrial uses. (Courtesy Proctor Historical Society.)

IMAGES
of America

VERMONT'S
MARBLE INDUSTRY

Catherine Miglorie

ARCADIA
PUBLISHING

Copyright © 2013 by Catherine Miglorie
ISBN 978-0-7385-9819-2

Published by Arcadia Publishing
Charleston, South Carolina

Printed in the United States of America

Library of Congress Control Number: 2012945463

For all general information, please contact Arcadia Publishing:
Telephone 843-853-2070
Fax 843-853-0044
E-mail sales@arcadiapublishing.com
For customer service and orders:
Toll-Free 1-888-313-2665

Visit us on the Internet at www.arcadiapublishing.com

For my daughters, Abigail and Bethany Miglorie.

CONTENTS

ACKNOWLEDGMENTS

I grew up in the town of Proctor, Vermont. Although my family did not work in Vermont's marble industry, the Vermont Marble Company was the center of life in Proctor in those days. My friends and I, many with parents, grandparents, aunts, and uncles who worked at the company, walked each day on sidewalks made of marble, over the marble bridge in the center of town, and past the Vermont Marble Company offices and shops to get to the elementary school at the top of the hill. History was all around us. This book is an effort to keep the history of the Vermont Marble Company alive for future generations.

My deepest thanks go to the Proctor Historical Society, housed at the Proctor Free Library, and especially Nancy Kennedy. All of the images used in this book are courtesy of the Proctor Historical Society's vast archives. I encourage all of you to visit the Vermont Room at the Proctor Free Library and flip through the extensive collection of scrapbooks, photographs, books, and files there to learn more about the history of the Vermont marble industry.

I began writing about Vermont's marble industry when I became affiliated with Dimensions of Marble, a nonprofit organization founded in 2005 to document and share the story of marble, its artistry, and most importantly, its people. One of our projects was a historical series, "Marble Minutes," which was published in the *Rutland Herald*. I was the second writer of that series. Please visit www.dimensionsofmarble.com for more information on the Vermont marble industry.

While this book is focused on the history of the Vermont Marble Company and Proctor, I want to acknowledge that the marble industry is an important part of the history of the towns of Dorset, Manchester, Rutland, West Rutland, Middlebury, Burlington, and other towns along Vermont's marble belt, on the western side of the state. It was not possible for me to include the complete history of the entire region in these pages.

Finally, thank you to my family for their love and encouragement on this project.

INTRODUCTION

In 1767, a man named John Sutherland journeyed up an old military road, a remnant of the Colonial and Indian wars, and staked a claim to the land below the great falls of Otter Creek. There, he built a sawmill and a gristmill. Soon, a little settlement arose named Sutherland Falls.

Marble quarrying had already begun in southern Vermont by 1785. By the early 1800s, several quarries had been opened in the West Rutland area. It was not until 1836 that Willard and Moses Humphrey began the first quarrying operations at Sutherland Falls. The marble industry started in that old mill down in the hollow, a building in which only six or eight quarry blocks could be sawed at a time. There was a track leading up to the quarry, and the blocks were placed on a car and lowered into the mill using a cable. A yoke of oxen hauled the cart back to the quarry again. Quarrying and milling methods were simple in those days.

After several attempts to make a viable business from marble quarrying, S.M. Dorr and J.J. Myers took over the operations in 1857, and it became known as the Sutherland Falls Marble Company. But financial woes plagued the marble company until 1869, when Col. Redfield Proctor took over the marble mills and quarries and reorganized the company into a financial success.

Colonel Proctor was a man of strength and determination, and under his control, the marble industry rapidly expanded throughout the 1870s. He absorbed the competing local marble companies into his business, forming the Vermont Marble Company. In 1884, the unincorporated village of Sutherland Falls officially became incorporated, and the town was named Proctor. The dynastic age of the marble industry had begun.

Orders for marble came into the Proctor offices from across the country. A new mill was erected to accommodate the growing business. Extra help had to be employed, and immigrants flooding into the United States created a readily available labor pool, sought after by Vermont Marble Company officials. New houses sprang up throughout the town. A company store, a post office, churches, and a school were built to meet the needs of the growing community. Foundations were laid for a library, a YMCA, and a hospital. In 1885, the Vermont Marble Company started the Clarendon and Pittsford railway, which was wholly owned by the company and connected the quarries in Proctor and West Rutland with the company mills.

Down at the marble yard in Proctor, a finishing shop was built to carve and polish the marble. A building shed housed the blocks and slabs used for filling the orders for block-style buildings, which were rapidly being built across the United States. More mills were added in Proctor, West Rutland, and Center Rutland so that quarry blocks could be more quickly cut to meet the demand for heavy-dimension stone. A machine shop and a woodworking shop were added. Every department was bending under the strain of keeping up with the nation's growing clamor for marble.

As demand for Vermont's pure marble grew, the rushing water of Sutherland Falls was harnessed for electric power, horses replaced oxen, and traveling cranes crept into the shops and yards. Other quarries—the Pittsford Valley, the West Rutland, and the Blue Rutland—were developed to supplement the work of the Sutherland Falls and Columbian quarries. The railroad was pushed to completion so that the Vermont Marble Company operations were seamlessly connected. A tramway was built to provide sand from a distant mountain to feed the hungry saws of the mills.

The marble company bought more land across the country, opening additional quarries as far away as Alaska. The marble from the Alaska quarries was primarily used for buildings in the Pacific Northwest, California, and the western states of Montana, Idaho, and Utah. Some of the prized Alaskan marble was shipped back east as far as Boston. The Vermont Marble Company purchased the Yule quarry in Marble, Colorado, whose pure white marble was used for the Tomb of the Unknown Soldier and the Lincoln Memorial. The acres of land on which the quarry holdings sat were pristine forestland, which the marble company harvested to turn into boxes for the finished marble.

When Redfield Proctor answered the call of Washington, DC, in 1889, Fletcher D. Proctor succeeded his father as president of the Vermont Marble Company. Under his direction, the company became the giant of the stone industry and one of the largest corporations in America. By the early 1900s, the Vermont Marble Company employed more than 4,000 men; had offices, quarries, and shops throughout the United States; and distributed one million cubic feet of marble around the globe annually. In 1912, the Vermont Marble Company stretched out along Otter Creek in Proctor in a chain of shops and mills with floor space of 25 acres, and had 10,000 horsepower of electrical energy on command, generated from its power station at the foot of Sutherland Falls. The buildings housed a score of turning lathes, more than 100 rubbing beds, 300 pneumatic tools, and more than 400 gang saws, running day and night.

In the early 1900s, at the company's height, 21,000 blocks were quarried annually, many of them being made into tombstones and monuments. Marble was used in the making of public buildings and private residences, libraries, and churches, and all manner of exterior and interior work. Redfield Proctor, who became the secretary of war and then a state senator from Vermont, used his political influence to ensure that marble from Vermont Marble Company quarries was used in many of the buildings and national monuments that grew to grace the Washington, DC, skyline. Buildings and memorials in our nation's capital, including the Arlington National Amphitheatre, the Supreme Court, the Thomas Jefferson Memorial, and the Tomb of the Unknown Soldier, were all created from the stone of the Vermont Marble Company's quarries and the working men of Vermont's marble industry. Because of Senator Proctor's political influence in the American business world, three US presidents visited the small town of Proctor as his guests.

Fletcher D. Proctor followed in his father's footsteps, serving as president of the Vermont Marble Company until his death in 1911. The Vermont Marble Company and the Proctor family carried great political influence in the state of Vermont. Fletcher D. Proctor, Redfield Proctor Jr., and Mortimer Proctor all served terms as governor of Vermont. Frank C. Partridge succeeded Fletcher Proctor as the president of the company. Under Partridge's tenure, the Vermont Marble Company had some of its largest growth.

The men who turned out the blocks of marble from the quarries and turned them into works of art comprised a melting pot of nations, each with their particular job in the machinery that made up the marble company operations. As the fledging marble industry took root, Irish and Canadian men were among the first to arrive, taking their places toiling in the quarries. Newly arrived Swedes were sought after by Vermont Marble Company executives, who sent agents to Ellis Island to lure the Swedes and other immigrant families to Proctor. The Swedish men were deemed reliable workers, and, unlike some employees of other ethnic backgrounds who were relegated to long hours in the quarries, Swedes found their way into all levels of marble industry work, from cutting and finishing marble to sales and managerial positions.

A majority of Norwegian settlers came to the United States by way of Canada, choosing to make their fortunes in the marble and granite quarries of Vermont. Finnish immigrants came here to answer the call of Vermont's railway companies, which sought men with experience on the rail systems of Northern Europe. Eastern Europeans, fleeing from countries like Poland and Belarus under the menacing shadow of Russia, were welcomed by the Vermont Marble Company and put to work in the company's mills and shops. People from the lands of Czechoslovakia, Austria-Hungary, and Germany left their countries en masse and found work in the quarries, mills, and stone shops of Vermont.

Northern Italian men were sought for their carving skills, while many Southern Italians immigrated to Vermont and became the backbone of the Vermont rail system in Rutland. The Rutland region became a rich tapestry of ethnic communities.

Although the Proctor family was known for their benevolence to their workers, life in the quarries and shops was not always easy. Men worked far below the ground in cold and damp conditions. Before the quarries were electrified, the equipment was run using steam piped into the quarries. Danger was a constant companion. In 1893, one of the worst accidents in the company's history happened when a large mass of stone broke loose and fell into the depths of the mine, burying a gang of men who were working the channeling machines. Seven men were killed and many others were injured. As soon as the accident happened, the electric danger signal was rung at the company office, and work everywhere, throughout all company departments, was stopped at once while rescue efforts were made.

The magnitude of the accident shook the industry. The marble bosses intensified their efforts to improve working conditions using modern technology. Changes were made to quarry operations. The steam pipes were removed and electricity was installed. This improvement took away much of the gloom and dampness and made working conditions safer and more comfortable.

The Proctor family was progressive, making every effort to ensure comfortable conditions in the town itself around 1900. The Vermont Marble Company built housing for its workers, which it rented to them, or sold them land at a very low price if they wanted to build their own homes. Proctor had a water system and electric lights for its 2,500 inhabitants and was a well-kept and up-to-date little town. The company provided the first industrial nurse in the nation, to care for its workers in their own homes, before establishing the Proctor Hospital. At the hospital, injured men were treated free of charge. Those who were ill were charged $4 per week, but only if they could afford to pay.

In the 1930s, the Vermont Marble Company was well established. Hundreds of buildings and thousands of memorial headstones were erected from the Vermont marble that was steadily churned out of the marble company factories. However, in 1936, during the Great Depression, even the mighty marble industry was affected. With revenues declining, working hours were cut, and employees faced the possibility of losing their jobs. More than 600 workers went out on a bitter strike. The men who stayed on the job in order to feed and house their families during the cold winter were stoned by the strikers as they reported to work each day. Their houses were bombed, power lines were blown up, and bridges were blown out. By July 1936, the marble bosses prevailed and the strike was broken. By 1945, the Vermont Marble Company was unionized.

The 1940s began a new era for the marble industry. During World War II, the Vermont Marble Company balanced the production of war materials with the creation of memorials in the Proctor monument shop. With men leaving for the armed services, women began taking their places at the machines. The marble plant in Proctor transformed a large portion of its operations for war purposes. The same planing machines that molded the massive marble sections for the National Gallery of Art and the Jefferson Memorial earlier in the century were now enlisted in the cause of war. Airplane parts, ship winches, and weapons were manufactured alongside marble memorials in the huge monument shop. For its wartime efforts, the Vermont Marble Company won the Army-Navy "E" award three times.

After World War II, the era of corporate America was ushered in. By the 1960s, demand for heavy stone blocks began to lessen. Instead, modern architects sought thin veneer panels for their building plans. The United Nations building, in New York, used veneer marble panels and was one of the last great building projects of the Vermont Marble Company. The company then focused on producing floor and wall tiles and searched for other new ways to use its marble. However, the heyday of the Vermont marble industry was over.

Business continued to decline, and in 1976, the Vermont Marble Company changed hands. It was purchased by the Swiss company Pluess-Stauffer, now named Omya, a mineral processing company. Omya is the world's largest producer of ground calcium carbonate. The company purchased the Vermont Marble Company in order to acquire its holdings, especially the prolific Middlebury quarry, where the stone was not suitable for dimension stone, or building stone, but was perfect for crushing into its purest form, calcium carbonate, for industrial uses. Omya still operates a calcium carbonate production plant in Florence, Vermont, as well as the Middlebury quarry, but moved its corporate headquarters from Proctor to Blue Ash, Kentucky, in 2007.

Today, the quarries and shops of the Vermont Marble Company in the middle of Proctor lie quiet and vacant. The town itself—the legacy of one man, Redfield Proctor, and his family—remains rich in history and still proudly exists amidst the glory of the surrounding landscape of Vermont marble.

One

FOUNDING FATHERS

John Sutherland started a sawmill and gristmill at the foot of the Otter Creek falls around 1767. Soon, marble was extracted from the first local quarries, and the Sutherland Falls Marble Company was born. This 1878 photograph shows the company office on the right, with the steeple. The big stone building in the center had a basement and three sub-basements, through which a direct rope driven by the waterpower of Sutherland Falls brought power to the sawmill above.

The mighty Sutherland Falls cascades approximately 118 feet down a tiered bank into a wide pool. First the site of John Sutherland's sawmills and gristmills, Sutherland Falls later became the source of hydroelectric power for the Vermont Marble Company and the town of Proctor. The power station, built in 1905–1906, continues to provide electrical power to the town today. The banks of the river above the falls were widened twice by blasting out stone. This alleviated water pressure during floods of Otter Creek and also provided additional water flow through the power station's turbines.

The Meads and the Humphreys were two of the original families that settled in the Sutherland Falls area. The Mead brothers, Zebulon, James, and Stephen, were instrumental in the settling of Rutland County—Zebulon at Sutherland Falls, James in Rutland, and Stephen in Pittsford. The Mead residence (above) was built in 1840 on West Proctor Road. The Humphreys were pioneers in the marble industry and opened some of the first quarries in Proctor. The Humphrey residence (below) was built in 1826. All the materials used in the construction of the home were provided by the land on which it was built, except the window glass.

Early quarrying methods were crude. At first, marble was simply pulled from outcroppings that were visible above the ground. Later, marble was blasted from the quarries with gunpowder in a wasteful process. The slabs that remained after blasting were pulled out of the quarry using a system of ropes, pulleys, and rollers up an inclined plane. They were then loaded onto carts or sleds and hauled to the mill at the falls by teams of oxen. The first mill had four gangs of saws and was built in the winter of 1836–1837. The first use of the marble was for cemetery headstones. It was not until the late 1800s that marble was used to any extent for other purposes. As quarrying methods became more refined, marble's popularity as a building material—for exterior and interior—grew rapidly. Eventually, sales for this purpose matched sales for cemetery memorials.

The operations in Sutherland Falls suffered during the financial panic of 1837–1838. From 1845 to 1854, operations at Sutherland Falls were virtually at a standstill. Competing quarries and mills had been opened in West Rutland and Center Rutland. In 1857, there was a company reorganization, and the Sutherland Falls Marble Company began. The new company's original buildings are seen here in 1878. Col. Redfield Proctor took receivership of the company from the financially challenged owners, S.M. Dorr and J.J. Myers. Soon, the building and the yards underwent extensive renovations, and the sign changed to the Vermont Marble Company as Redfield Proctor began to grow his marble business.

Before the Vermont Marble Company expanded operations, the company was run from the original old mill. Only six or eight blocks were sawed at a time from the old quarry, and then run by teams and carts to the mill. For many years, horses and oxen were an integral part of the marble company. In 1849, the Rutland & Burlington Railroad was completed. With the advent of the railroad, transportation methods for both people and stone changed dramatically. The residents of the little village were no longer cut off from the rest of the area. By the 1870s, rail spread throughout the region and into the quarries themselves to haul the marble ore to its final destination.

In 1869, Col. Redfield Proctor took control and renamed the operation the Sutherland Falls Marble Company. Colonel Proctor had a vision for the marble company. With its abundant waterpower, promising quarries, a good deposit of sand on the site of the present marble yard, and the railroad at hand, he believed the company could prosper using sound and efficient business practices. Over the next 10 years, Redfield Proctor personally and untiringly loaded cars, selected and graded marble, and oversaw all the details to build the company into a prosperous business. By 1880, the company operated 64 gang saws and was very successful.

SENATOR PROCTOR

Redfield Proctor was an astute businessman, focusing only on the marble business, in which he made sound investments. He reinvested all profits into the business's growth and used his signature traits of concentration, energy, singleness of purpose, organization, and economy to make the marble industry flourish. Colonel Proctor was also an avid hunter in his leisure time.

In 1889, Redfield Proctor began his political career as the secretary of war under Pres. Benjamin Harrison. This picture of Harrison's cabinet shows, clockwise from left, John Noble, the secretary of the interior; John Wanamaker, the postmaster general; Jeremiah Rusk, the secretary of agriculture; William Miller, the attorney general; Proctor; William Windom, the secretary of the treasury; James Blaine, the secretary of state; and President Harrison.

This photograph shows Redfield Proctor's house decorated for Pres. Benjamin Harrison's visit to Proctor in 1891, when Proctor was the secretary of war. In Harrison's speech to the residents of Proctor on this visit, he hinted of an upcoming change in his administration; in November, Harrison appointed Proctor as a US senator. Proctor was subsequently reelected to the senate, serving from 1892 until his death in 1908.

In August 1897, Pres. William McKinley visited the town of Proctor. Senator Proctor's home was decorated with a formidable barrier of shrubbery decorated with red, white, and blue incandescent lights and a glowing fountain. More than 1,000 people welcomed the president, first lady Ida McKinley, and the secretary of war, Gen. Russell Alger.

Proctor's claim to fame as the Vermont town most visited by presidents continued on September 1, 1902, when Pres. Theodore Roosevelt arrived in Proctor on a special train. The railway station was patriotically decorated. The station was freshly painted for the first time since it was built, and a special hemlock hedge was planted overnight to hide an ugly fence along the route of the parade.

This photograph shows the west entrance of the covered bridge spanning Otter Creek at the time of President Roosevelt's visit in 1902. Because Roosevelt's visit was so short, only this end of the bridge—the only part visible from Proctor's home—was decorated and given a whitewash bath.

On September 1, 1902, Pres. Theodore Roosevelt addressed a large crowd from the piazza of Proctor's home. He arrived in Proctor at 11:50 a.m. and was greeted by the Proctor Band. Roosevelt began his speech: "It is a very great pleasure to be here and greet you from the home of my old and valued friend." He spoke at length about the Monroe Doctrine, establishing the basis for what would come to be known as the Roosevelt Corollary to the Monroe Doctrine in 1904, in which he warned European forces to keep from intervening in matters of power in the Americas. After his address, he departed at 12:30 p.m. from the Proctor train station to Rutland, where he also addressed a large crowd. In what was now a tradition, Proctor's home was ornately decorated with patriotic bunting and shrubbery.

Senator Proctor died on March 4, 1908. These photographs show the arrival of the special train bringing his remains to his hometown of Proctor. Funeral services were held at the Union Church. His pallbearers were his sons Redfield Proctor Jr. and Fletcher D. Proctor, the governor of Vermont; E.F. Holbrook, the senator's secretary; S.A. Howard, a family friend; F.C. Partridge; and B.F. Taylor. The funeral was attended by an estimated 10,000 people. Approximately 3,000 company employees lined the half-mile route from the Union Church to the family mausoleum in the Proctor Cemetery. Senator Proctor was revered by townspeople as the founder of the Vermont Marble Company, although he had handed over the role of company president to his son Fletcher at the start of his political career.

Two

THE CALL OF THE QUARRIES

The richest marble deposits in Vermont lay between Proctor, Middlebury to the north, and Dorset to the south. Other Vermont quarries included ones in Swanton, which produced a reddish marble; in Rochester, which produced Verde Antique green marble; and in the Burlington region, which was home to Lake Champlain Black marble. In Proctor alone, 31 quarries were mined during the peak years of the marble industry. The Sutherland Falls quarry, seen here, was one of the most prolific of the Proctor quarries.

MARBLE QUARRY Proctor. Vt.

The first quarry in Proctor, the Columbian, was opened by Willard and Moses Humphrey a short distance east of West Street in 1936. At the time, the men had little capital to pursue the venture and little knowledge of the magnitude of the work they were undertaking. They partnered with Edgar L. Ormsbee of Rutland under the company name of Humphrey and Ormsbee, and built the first mill in 1837. A financial crisis gripped the country in the winter of 1837–1838, and the firm went under, granting all rights to its creditors. The business was then assigned to Francis Slason of West Rutland. Moses Humphrey remained the foreman of the mill for several years. This 1910 photograph shows the Columbian quarry, which was behind (east of) the Tenney Humphrey house on West Street. After opening this quarry, they began prospecting for additional marble deposits, finding the Sutherland Falls deposit in 1838.

The Columbian deposit was the most valuable of the Proctor quarries. In 1838, the Humphreys opened the Sutherland Falls quarry north of Market Street, which was operated on and off by the Vermont Marble Company until April 11, 1929. The Sutherland Falls quarry and the Columbian quarry were two large veins of marble separated by a mass of dolomite that measured about 150 feet in width. The first work at the Colombian quarry was done by blasting out the stone with gunpowder. The stone was then hauled out of the quarry using a system of ropes, pulleys, and rollers up the bank to a team of oxen, who dragged them to the mill near Sutherland Falls. In the course of quarrying the Colombian, several smaller digs were made in the adjoining mass of dolomite. Soon, they found the marble belt that became the mighty Sutherland Falls quarry.

The channeling machine was the workhorse of the quarries. Invented by George Jeffards Wardwell in 1863, the first machines were steam-operated. Drills were set into frames and the steam power moved them up and down through layers of rock. In later years, the channeling machines were operated by electricity. The Sutherland Falls Marble Company commissioned the first channeling machine from Wardwell. They agreed to pay for the construction of the machine, and Wardwell received $2.50 per day for supervising its construction and testing the machine's suitability in the Proctor quarry. In 1865, Wardwell sold the entire interest in his patent to the Steam-Stone Cutter Company, receiving $1,500 in cash, 3,352 shares of stock, and a role as the superintendent of the company. By 1868, the Steam-Stone Cutter Company had built and sold 24 machines. One of these was sent to the Paris Exposition of 1867, where it won a silver medal.

Before the channeling machine's invention, marble was wastefully blasted from quarry walls or chipped out with hand drills by teams of men in a time-consuming process. Quarry owners quickly saw that the old, labor-intensive methods of stone extraction cut deeply into the profits of the growing industry. Innovation was needed, and what became the most widely used of all quarrying equipment, the channeling machine, was invented in 1863. The channeling machine worked by tunneling through layers of rock to drill grooves into the stone walls. The blocks were lifted using "plugs and feathers"—wedges and shims—and released from their beds.

These channeling machines at work in the Sutherland Falls quarry were supplied by the Sullivan Machine Company of Claremont, New Hampshire, and the Steam-Stone Cutter Company of Rutland. The channeling machine had rows of long chisels set in a strong traveling framework. This gang of chisels was powered by machinery and vibrated up and down, cutting a channel or groove in any desired direction. When the groove was sufficiently long and deep, the machine was moved to another place to cut a cross channel. The bottom was also perforated. The block was then easily split away using wedges. The blocks of marble were lifted by cranes and derricks worked by steam or electricity and then carried rapidly to the railroad cars for transportation.

In 1870, when this photograph was taken, the Sutherland Falls quarry was barely 20 feet deep. Men went into and out of it using ladders. A pulley system was used to lift blocks, which were then loaded onto carts and transported by a team of oxen to the mill for processing. At this time, the Sutherland Falls Marble Company, a Massachusetts company, operated the quarry and a mill consisting of 10 gang saws. When Colonel Proctor took over the company, he reorganized the Sutherland Falls Marble Company into a Vermont corporation. The Sutherland Falls quarry, with its seemingly unending supply of pure marble, was the foundation of the business's success. As the men dug deeper into the earth, quarrying methods were refined to use the electrical power harnessed from Otter Creek to loosen the blocks from their walls of stone.

This is a view looking into the Sutherland Falls quarry. Both ladders and stairways were used to access the quarry. Four channeling machines were used to drill through the hard marble walls to release the slabs of marble. Core drills were used to bore into the marble, testing the soundness and quality of the block.

At the top of the Sutherland Falls quarry is a steel railway bridge that runs out across the quarry. Wire cables are attached to the bridge, which rests on large blocks of marble. The marble blocks are set onto the railcars and transported to the shops for processing.

Workers dug ever deeper into the earth as the quarries yielded their rich ore. Stairs led down into the Sutherland Falls quarry. The work in the quarry was physical and dangerous, and men endured their share of accidents and tragedies, including death. Facing danger on a daily basis was simply part of the job. Falling marble slabs, equipment failure, and electric shock were among the most common causes of accidents in the industry. Company officials soon realized that in cutting marble out of the ground, it was not enough to get it out quickly and in block sizes that met the demands of the mills and shops. It was also important to get the stone out safely. The men began to systematically cut the stone from the wall to eliminate scaling and undermining, which could affect the quarry's internal structure. Any collapse of the quarry structure caused an extended delay to the smooth running of the marble business.

ONE OF VERMONT MARBLE CO'S QUARRIES PROCTOR. VT.

This postcard shows the Sutherland Falls quarry operations from inside the quarry. After the invention of the channeling machine, quarrying methods were adapted to fit the machine. After much trial and error, the uneven working surfaces typical of a quarry were brought down to level floors. Then, the machine was able to tunnel through the layers of rock. The depth of the quarry here is more than 200 feet. The Sutherland Falls quarry eventually reached a depth of 300 feet. The marble it produced was very strong and durable, due to its close graining, and was composed almost equally of carbon dioxide and calcium oxide, with very few impurities. Its white color was slightly cloudy and variegated. The color and the strength of the stone made it well suited for large blocks for exterior buildings.

This early view of the Sutherland Falls quarry was taken in the 1870s. The building in the photograph is the crane shed, which was used to store machinery for handling and lifting the marble blocks. The crane shed was built at a cost of $40,000, and was considered to be an unwise use of funds. Another project deemed a financial mistake was the building of a pipe to carry water from Beaver Pond to the quarry in order to propel the hoisting machines. These projects were blamed in part for the company's subsequent heavy financial losses, which forced the Sutherland Falls Marble Company to be placed in the hands of a receiver, Col. Redfield Proctor. The new owner also believed in investing the company's profits back into the business, but his capital investments were much more successful than those of the previous owners.

At one time, the Vermont Marble Company owned quarries from Alaska to Vermont. In Proctor alone, 31 quarries were opened to fuel the mighty company's need for stone. In 1901, the company had possession of the Verde Antique deposit in Roxbury and the Champlain marble deposit in Swanton. Purchases at Danby began in 1905. In 1909, it acquired the property of the Brandon Italian Marble Company, which included quarries in Brandon and a plant in Middlebury. Before the Vermont Marble Company purchased the West Rutland deposits, 15 different companies operated its 21 quarries at one time or another. For its first 75 years of operation, the greater portion of the production from these quarries was used for headstones. All records of those early years refer to amounts produced as "thousands of feet, superficial measure, two inches in thickness," the standard thickness for cemetery memorials in the mid-1800s.

This view looks into the Sutherland Falls quarry. Over time, the quarry became much deeper. The Sutherland Falls quarry was the main producer of marble for the mills and shops of Proctor for many years. Stone was taken from the quarry, loaded onto block cars, and transported by train directly inside the mills. The train tracks from the quarry to the mill were laid in 1871. Because of its strength, soundness, and white-gray color, the stone from the Sutherland Falls quarry was used in a number of notable buildings across the country, including the post office and courthouse in Montpelier, Vermont; the spire of Grace Church in New York; the Second National Bank in Paterson, New Jersey; and Clio Hall at Princeton University. It was also used extensively throughout the town of Proctor in its many marble buildings.

Because of the unusual positioning of the marble beds, quarry openings were somewhat like the shafts of a mine, leading down into tunnels, which were the active producing centers. The great excavation at West Rutland started as an open-faced quarry, but after reaching a depth of 200 feet, the marble vein turned abruptly back under the hill. The only way to get the marble was to follow the vein. The first attempt to quarry marble from this deposit took place about 1833. The first operator was William F. Barnes, whose father, it was stated, purchased almost the entire deposit, giving in exchange an old horse worth $75. Barnes was a man of many abilities, which served him well in developing his own marble properties. He reportedly realized more than $130,000 on his father's purchase before he was killed in one of his own quarries by a piece of marble that fell from the surface, hitting him on the head.

The development of the West Blue, commonly called the Harrington quarry, began in 1901. Above, men are at the bottom of the Alberston quarry in West Rutland. A string of electric block cars traveled along this electric road, or railway, inside the West Rutland quarry. In 1923, the block cars climbed 500 feet to the top of the quarry with loads of heavy blocks. At the time, the West Rutland quarry was a combined stretch of tunnel covering a distance of about 2,000 feet. It was hundreds of feet wide and interlaced with lines of electric car track.

In the West Rutland covered quarry, stone pillars were used to support the roofs of the rooms, or system of tunnels, from which the stone was extracted in blocks. On the roofs of these tunnels rested the weight of an entire hill. These supports, along with careful pointing of the intervening roof spaces, prevented the quarry from collapsing. After quarrying, the marble was raised 900 feet to the surface using an inclined railway system.

About 500 men worked in the vast underground quarry in West Rutland in 1923. Illuminated by strings of electric lights, men worked the rattling drills and channeling machines to bore into the massive marble walls as they struggled to release its massive load of stone. Other gangs of men separated the waste marble from the desirable slabs and loaded them onto the inclined cable track. West Rutland quarries had layers, ranging from a pure white marble to an almost solid green, and the West Blue layers were varying shades of gray with clouds and veins of darker shades of the same colors.

Descending a series of stair flights was a common way to get down to the work areas of Vermont quarries. The steps could number several hundred in the course of the descent. Here, in West Rutland, a slab of marble is hauled by crane onto the quarry surface. A mammoth eight-sided pier supports the outer walls of the quarry. These piers were often used to support quarry walls. Some were carved out of the very stone of the quarry wall. Others were built using reinforced concrete. These jut out between the east and west machine-made marble cliffs like the girders of a huge bridge. Since the cavity of the quarry increased in length and breadth as it grew deeper, each successive pier was correspondingly longer than the one above it. It was quite a feat to mold the octagonal beams and pour the concrete in midair during construction.

Three

FROM EAST TO WEST

Several large quarries outside of the Rutland region provided stone for monuments and buildings on the west coast and in Washington, DC. The Danby quarry, seen here, still operates today. Stone from the quarry in Tokeen, Alaska, was mined and finished in Tacoma, Washington, and San Francisco. The Yule quarry's claim to fame was providing stone for the Tomb of the Unknown Soldier and the Lincoln Memorial.

The photograph above shows the entrance of the Danby quarry in 1945, leading into an underground maze that twisted and turned deep into the mountainside quarry. Since the late 1700s, men labored to pull the prized marble from deep within the Danby mountain quarry, inventing innovative and laborsaving quarrying methods along the way. The 80-ton block below would be hoisted from the quarry entrance by crane to the bank to be loaded onto the cable railway, where it would be sent down the hill to the railroad station 900 feet below the quarry entrance.

The massive derrick swings the block from the Danby quarry entrance and sets it on the bank. There was a time when every block of marble that came off the mountain had to be loaded on a wagon and sent down over a treacherous, winding road. In the rainy seasons, the way was often blocked by mud, or by banks of snow in the winter. When the derrick and cable car system was invented, it shortened the removal time.

In this 1945 photograph, a man maneuvers the 80-ton block onto its resting place on the bank, hundreds of feet above the train tracks. As quarrying methods progressed, blocks were hardly out of the Danby quarry before they were set down in the valley, where railcars awaited to bring the blocks to the Vermont Marble Company shops for finishing.

This photograph shows the top of the cable road, which was a modern marvel when it was built. It was dependable, running in all kinds of weather—mud and snow were no longer a menace. The track was a mile long and caused many problems with its construction far up the hill to the mountain quarry. It was anchored by a powerhouse at the top of the road that contained a mammoth drum, around which the wire rope was wound. When the operator turned on the power, the car at the top of the line started to go down while the car at the bottom started to go up. A stretch of double track provided a meeting place. The operator watched the indicator and stopped the wheels when the car came to the end of the track.

The cable road trailed down from the upper quarry down a track more than a mile in length. There was a long stretch of wire with a car attached to either end. The block was transported down the mountainside to the waiting railcars at the tracks below. Creaking with the weight, the cable car, with its load of an 80-ton block, rode slowly down the steep track of the cable road. Once at the bottom, another team of men sprang to work, manning the derrick and maneuvering the heavy block as it was lifted up from the cable car. The derrick winched the huge block of stone gently onto the waiting railcar. The men strapped it into place and gave the signal for the train to depart for the mills and finishing shops in Proctor or Center Rutland.

Vermont quarrymen pose after raising this 80-ton block from the Danby quarry. Their attitudes of nonchalance belie the tremendous effort made in dealing with that great mass of stone. Their task was of monumental proportions: raising the massive block from the quarry onto the bank, loading it on the cable car, sliding it down the steep hillside, and finally loading it by block and tackle onto a regular railroad flat car for its journey miles north to the mills of the marble company. In the photograph above, the huge cable from the incline railroad is in the foreground.

This 1945 photograph shows the interior of the Danby quarry, which is a maze of many tunnels. At that time, the quarry was hundreds of feet wide and interlaced with lines of electric car track. Its mighty roof was supported by gigantic marble piers, which have been left intact, while the blocks around them have been cut away. Today, the entrance to the Danby quarry looks the same as it did a century ago. However, the quarry is now more than a mile wide and encompasses 25 acres deep inside Dorset Mountain. The large quarry cavity is known as the Imperial quarry and is the source of the company's most popular marble. A second level, the, Brook quarry, is also active. A third level, the New Imperial quarry, was started in 2006. The work is done by a staff of less than 30 people, local Vermonters, plus a couple of veteran Italian quarrymen.

In the Danby quarry, men spray walls in preparation for drilling. Here and there in the impenetrable gloom of its deepest levels are clusters of blinking lights, illuminating the men's work. The Danby quarry continues to be mined today, using traditional quarrying methods, although the blocks are no longer transported via the inclined railway. In 2002, a full-scale production facility, complete with saws and polishing equipment, was added to the quarry inside Dorset Mountain. The blocks are milled right inside the massive quarry, and up to 4,000 square feet of slabs per day can be produced. The Danby Quarry was the source of the marble for many buildings and monuments in Washington, DC. It also supplied stone for many modern projects, including the United Nations building in New York, the Knowlton School of Architecture at Ohio State University, and the new additions for the Montreal Museum of Fine Art, the Art Institute of Detroit, and the Museum of Modern Art in New York.

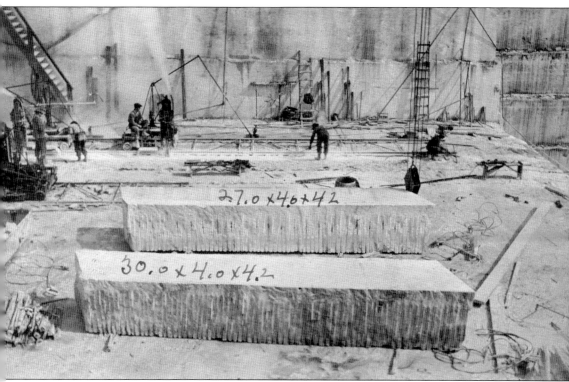

In 1908, the Vermont Marble Company started quarrying operations in Tokeen, on Marble Island, Alaska. Long ago, Southeast Alaska Indian children played with dolls that had marble heads, chiseled from the marble from the Tokeen deposits. The Alaskan deposits were not as stratified as the ones in Vermont, but instead lay in large irregular masses. Many openings were required to stabilize the output of the stone. The deepest Alaskan quarry was about 160 feet in 1928. Blocks of marble, some weighing as much as 11 tons, were quarried from the island and loaded onto flat cars and then barges for shipment to the Vermont Marble Company mill in Tacoma, Washington. The quarrying season covered nine months, between February and December. At the peak of marble production, there were eight quarries employing 70 workers at Tokeen. The workers lived in bunkhouses, took their meals in the company dining hall, and worked six days a week for those nine months of the year.

Boats and barges were integral to the Alaska quarrying operations. Marble blocks were loaded onto cars and run by gravity from the Tokeen quarry. The gravity-operated cars traveled down narrow-gauge tracks to the wharf. There, they were transferred by derrick to barges and shipped to Tacoma, Washington, or San Francisco, where the Vermont Marble Company had established shops and mills for extensive finishing and sales all along the west coast. The barges had a capacity of about 800 tons. The blocks ranged in weight from six to 20 tons. This photograph shows blocks being unloaded in San Francisco. The Vermont Marble Company owned four barges, which not only moved marble, but also men and even mess houses. In the early 1920s, two mess houses were moved 30 miles on rafts being towed around Marble Island by boats. They eventually landed at the new base of operations. The Tokeen quarries yielded more than $2.5 million in marble before the operation ended in 1927.

The 1927 photograph above of the Vermont Marble Company ship *Vermarco* shows blocks being shipped from Tokeen. The ships were also used to bring men to Alaska to work in the quarries. Hiring an adequate number of laborers was a recurring problem for the company. Of the men who agreed to go, some never even came on board in Seattle. Others would reach Tokeen but stay only long enough to get the next returning boat. The Tokeen operation was isolated from the rest of the world, and this isolation was the main reason workers left. Radio broadcasts brought Tokeen into closer touch with civilization, but the radio could not be turned on while the electric light generator was running, and the generator ran for power in the quarries until 9:30 p.m. each evening.

TOWN OF MARBLE, COLORADO

The Yule quarry, outside of the town of Marble, Colorado, is known for its marble blocks of pristine white. The quarry was first mined in 1873 and went through a number of owners before the Vermont Marble Company purchased it in the early 1900s. This 1932 photograph shows the town just after the 56-ton block of marble for the Tomb of the Unknown Soldier was extracted in February 1931. Yule was famous for producing the whitest marble in the United Sates, second only in the world to the stone of Carrara, Italy. The stone is 99.5 percent pure calcite, with a grain structure that gives it a smooth texture and a luminous surface. The Yule quarry houses extensive marble deposits, and very large blocks could be quarried out of it, which is one of the reasons why the marble for the Tomb of the Unknown Soldier, with its 56-ton die block, was quarried from Yule marble.

Above is an interior view of the Yule quarry, looking down into it and northwards. Pulling out blocks of this size in any quarry involved a vast amount of intensive preparatory work, because the quarry is situated high up in an almost inaccessible mountain range. Yule quarrymen were known for their poise and calm when dealing with the huge blocks, averaging in size from eight to ten tons. They raised them from the quarry floor in sure precision and placed them carefully on a rail flat car or on the quarry edge. Below, two men operate steam and electric drills on a marble ledge. Above them on another ledge, channeling machines work to release other blocks.

This is the view from above the Yule quarry shop looking north towards the cable tower and derrick. The boom leaned out over a quarry opening, dropping a thick cable from its head to the depths of the quarry. On the quarry floor 125 feet below, a great white marble block was attached to the cable, slid an inch or two, and then lifted towards the daylight.

Quarrymen pose with the 56-ton block of marble taken from the Yule quarry in 1931. At that time, the block was believed to be the largest ever quarried. It was loaded on a railcar to be shipped across the country, bound for the Proctor shops of the Vermont Marble Company. There, it was carved into the sarcophagus of the Tomb of the Unknown Soldier, and made its final journey to Arlington National Cemetery in Virginia.

Four

RISE OF A GIANT

This photograph from the late 1800s shows the company after Redfield Proctor took control. In addition to monument work, Proctor started operations to process marble for exterior finishing. It began in a small way, with contractors coming to the mill and cutting marble furnished by the marble company. In 1880, an experienced foreman, Thomas Hagan, was hired to expand exterior building material operations.

In the early 1900s, the operations of the Vermont Marble Company were in full swing. Seen above are the Proctor shop yards, anchored by the home office building, with the mills in the background. Immigrant men were recruited to work in the marble industry when they first set foot on Ellis Island in New York. They came to Vermont to make a new life. Often, men of the same nationality worked together in groups performing the same type of job. Below, men pose for the photographer in front of a marble mill at the Proctor plant in 1905.

Stretched out beside Otter Creek was a chain of shops and mills, with a floor space of 25 acres and 10,000 horsepower of electrical energy at its command. In 1908, white, light blue, and gray-tinted marble was quarried and used in cemeteries and for the interior decoration of houses and large buildings. Shipments amounted to 8,000 carloads annually, valued at $3.5 million. Payroll was $130,000 a month, the yearly output of the quarries was 21,000 blocks, and the shop yards held an inventory of 20,000 blocks. In the early 1900s, the Vermont Marble Company employed approximately 4,000 people.

This 1927 photograph shows the construction of the new monument shop at the Proctor location of the Vermont Marble Company. The large, modern shop was a huge structure of steel, concrete, and glass. It was a complete plant, with sawed marble going in at one end and coming out at the other end boxed and ready for shipment. Dynamite was used to topple the walls of the old mill, which stood in the way of the new shop. It marked the close of the second span in the history of Vermont Marble memorials, an 80-year period. The old building was razed in order to make the new shop, but the 1868 foundation of the original building remained as the cornerstone. Construction was done by the Lackawanna Steel Construction Corporation. The 400,000-square-foot mill building opened in April 1928.

This aerial photograph was taken sometime after 1928. The new, large monument shop is shown, and the new, marble ashlar company office building is across the street from the company shops and yard. Railroad tracks crisscross the yard. Behind the main buildings (at left) is a storage yard of marble slabs adjacent to Sutherland Falls. The Proctor family home sits on the rise above the falls.

The new Vermont Marble Company headquarters was built in 1924. It showcased the use of marble ashlar, which was once considered waste stone, in its construction. In the 1920s, driven by a need for economy and further profit from the marble business, the Vermont Marble Company started marketing and selling the rejected marble and its variety of texture and color for ashlar, or square blocks of building stone, which became a popular choice for buildings.

rmont Marble Company - Proctor Plant.

Redfield Proctor followed industry-best practices as he began his marble business. After the Civil War, it was common for skilled craftsmen to handcraft products from start to finish. Often, they worked alongside the business owner. Proctor was a well-known figure at the quarries and the shops, working next to his men. As the marble company grew, the nature of labor changed with the Industrial Revolution. The company started using new, mass-production methods to fulfill the demand for pure Vermont marble. Workers became responsible for only a small part of the

process, performing one specific task repeatedly. These jobs required little skill, and Proctor hired immigrants who were willing to work for low wages to live and work in his town. While factories across America became impersonal environments where the pace of work was set by the capabilities of the machines, Proctor continued to be a fair and equitable boss. The standard of living in Proctor was well above the nation's norm.

This 1930 photograph shows the modernized Vermont Marble Company buildings. At this point, the marble company was producing finished marble for iconic buildings and monuments like the Tomb of the Unknown Soldier and the Jefferson Memorial as well as the lucrative production and sale of marble headstones. In 1934, The Vermont Marble Company opened the Marble Exhibit in the old monument shop.

The Marble Exhibit was originally intended as a showroom to display the varieties of marble available for interiors, mantels, floor tiles, monuments, and headstones to architects and builders. Soon, the Vermont Marble Company began to welcome visitors to the showroom. In 1936 alone, more than 30,000 visitors, from every state in the country and 24 foreign countries, visited the Marble Exhibit. This photograph shows a vintage quarry drill that was part of the exhibit's collection.

The Vermont Marble Company built a powerhouse at the base of Sutherland Falls to harness the power of the water cascading 185 feet down the falls of Otter Creek. The powerhouse, built in 1904, had an electrical generating capacity of more than 3,000 horsepower. The electrification of Sutherland Falls was made in 1904–1906. The company also bought and electrically developed Huntington Falls in 1910. Belden Falls was purchased in 1904 and electrified in 1913. These three great hydropower facilities tied together the company's various Vermont quarries and mills with 73 miles of high-voltage lines.

These photographs of the power plant were taken in the 1940s. The power station provided electricity to the entire town of Proctor at very low rates, and the facility was of great interest to the townspeople. The Vermont Marble Company Power Division was owned privately and was not a public-service corporation. Since it did not enjoy rights of eminent domain, the superintendent of real estate for the Marble Company, Mr. Boyce, had to secure rights-of-way by signing contracts with local property owners for the entire 73 miles of power lines between 1904 and 1913. In 2011, the power division was sold by Omya, which purchased it as part of the 1976 Vermont Marble Company deal, to Central Vermont Public Service Corporation (now Green Mountain Power).

Proctor was ravaged by several floods over the years. The most devastating occurred in November 1927. The dramatic photograph above shows the sand tramway buckets at rest during the flood. Operations at the marble company ceased as all efforts went to fight the floodwaters. The damage was extensive. The foundations of the old monument shop were badly undermined and much of the machinery was wrecked. At the building shop, the destruction was restricted to certain sections. The yards on all sides of the shops were gutted, and piles of marble were deeply buried in the debris. The hydroelectric station was a mass of sand and wreckage, shutting off all lights and power for the entire town. At the flood's peak, it was thought that even the marble bridge (below) might give way under the strain, but it fortunately remained standing. The company's new main office, which stands on considerably higher ground, was left untouched.

The climax of the flood at Proctor was during the night of November 3, 1927. At that time and during the following day, the river was at record height. The raging waters surged down through the railroad cut into the village square, out across the marble yards, then on through the rocky railway gorge to the north, scooping out a huge chasm on their way to the lower level. Three large tenement houses toppled into this treacherous pit, and the families rushed out into the darkness, in most cases saving little except the clothing they wore. Nine families lost their homes completely, and many others whose houses were underwater were left temporarily without shelter. All these people were given food and lodging in the new high school, and relief efforts were coordinated by local Red Cross members and their helpers. There was no loss of life in Proctor, and only one person at outside plants, John Cebula of Center Rutland, was reported dead.

In the early days of the marble mills, a few teams of horses could haul all the sand that was needed to feed the hungry saws from the nearby sandpit. The first sand removal was all done by hand labor. Men filled wagons, and horses were used to draw sand to the mill. As marble company operations grew, a tramway was built to haul sand two-and-a-quarter miles from the sandpit through a system of buckets and cables. The Vermont Marble Company sandpit was between four roads in the north end of Rutland. The 33-acre sandpit was bounded on the west by North Grove Street, on the south by Cedar Avenue, and on the north by Pinnacle Ridge Avenue. Buckets were filled in the sandpit and made their way over East Mountain back to the mills in Proctor.

In this 1921 photograph of the Vermont Marble Company, the sand tramway stretches over East Mountain. Marble sawing required immense amounts of sand to fuel the iron saws. Sand and water were mixed together to form an abrasive, which aided the saw in cutting the marble.

This photograph shows men at the Rutland sandpit. Each bucket was filled with 500 pounds and moved along the cable tramway at a rate of one every 50 seconds—or 72 buckets per hour—to the busy mills in Proctor.

Trains wait at the Vermont Marble Company loading dock in Proctor. With the advent of the railway and the invention of the electric-powered crane, heavier blocks of marble could be moved more quickly.

At the Proctor yard, men pose with blocks of marble that are ready to be shipped. The Clarendon & Pittsford No. 9 engine waits in the background. The Rutland region prospered as various industries swiftly sprang up to fulfill the needs of the marble trade.

Loco. No. 8 • Florence, Vt. • 1922 •
L. to R. S. Nelson, Brakeman, P. Breshnahan, Conductor;
C. Peterson, Engineer; C. Stanley, Fireman; H. Nelson, Brakeman.

The Vermont Marble Company required massive amounts of raw materials and other goods to keep operations running, all of which were brought to the company by rail. By 1923, a total of 100 tons of steel were needed every year simply to make the drills used in the channeling machines and gadders. The mills require about 23 cars of saw blades, 90,000 in all, or about 300 per year. In addition, 25 rubbing bedplates were needed to keep the shops going, with a combined weight of 255 tons. In a year, the company bought 300,000 bolts and about four million nails. Coal was needed for supplementary power requirements, for the heating of the various plants, and for purchase by townspeople in the cooperative stores, to the tune of an annual total of 12,000 tons, or about a carload a day.

The Vermont Marble Company electrified its operations in 1905 and 1906. Using traveling cranes, operators were able to offload the large blocks of marble from the railcars as they came from the quarry. They also used cranes inside the shops to easily move the blocks during processing. The hook-and-line mechanism of the crane ran along a horizontal beam that itself ran along two widely separated rails. In the long factory buildings that made up the Vermont Marble Company, the crane ran along rails against the building's two long walls.

The original mill at the West Rutland location was built in 1870, when the quarry operation was owned by the Rutland Marble Company, before the Vermont Marble Company's acquisition of the West Rutland operations. While the Proctor shops were the Vermont Marble Company's base of operations, the company also built an extensive addition to the original mill next to its West Rutland quarry operations. About 450 men were employed there in the late 1800s. This photograph shows the marble yard and mills. These mills were the first stop for many famous monuments, such as the 56-ton block for the Tomb of the Unknown Soldier on its journey from the Yule, Colorado, quarry. Today, the old West Rutland manufacturing facility is home to The Carving Studio, a nonprofit educational organization that holds year-round sculpture classes, workshops, residencies, and exhibitions.

As the great extent of the deposits of the West Rutland quarry was realized, shops and mills sprung up beside the quarry to begin the finishing process of the West Rutland stone. The mill at West Rutland was built by the Rutland Marble Company in 1870. The first building was a steam mill with 16 gang saws. In 1881, Sheldon & Sons added a 20-gang mill with the capacity for 48 gangs. All of these mills ran night and day. Their finishing department was added in 1879–1880, employing at times 120 men to make stock for the trade. Eight turning lathes, six polishing lathes, and three rubbing beds were in use.

In 1835, William Ripley and William Barnes, who foresaw how profitable the marble business would be to the region, purchased the waterpower rights to Otter Creek in Center Rutland and erected the first mill at the site. As the marble business grew, so did the mills, in order to meet the increased demand for marble slabs. In 1877, one of the mills was changed to a finishing shop and a turning shop. The Vermont Marble Company further expanded the plant on Otter Creek, south of Proctor in Center Rutland, upon its acquisition of the property. The original eight-gang sawmill was replaced by a 14-gang mill in 1882. This plant was primarily a mill, where stone was sawed and rubbed before being sent to Proctor for final finishing and carving. It was powered by the Belden Falls hydroelectric facility, which was owned by the Vermont Marble Company.

Five

ALL IN A DAY'S WORK

There were many types of jobs available at the Vermont Marble Company. Once the marble blocks arrived at the shops, there was much work to be done. The marble was sawed, polished, turned on lathes, and machined before it was given to the sculptors. It had to be graded and boxed for shipment. Men in the office created the shipping documents. Traveling salesmen completed the workforce, selling finished marble tombstones and building materials across the country.

Above, men inspect a great slab of milled marble in the Proctor shop yard. After milling, the marble was graded and marked for the finishing shops. The blocks were marked with height, width, and weight and then transported into the finishing areas by crane and train. Inside and outside, the railway was integral to company operations. Trains were used to transport marble blocks from the quarry. Tracks led right into the mills to drop their heavy loads. Below, a locomotive in need of maintenance is worked on inside a repair shop in the mill.

The Vermont Marble Company sawing mill used an age-old method of sawing stone. Smooth iron bands set in a frame were fed by sand and water and moved back and forth across the marble by power-driven machinery. Marble has a well-defined grain in one direction, but it can be sawed in any direction across the grain. Sawed marble produced a smoother finish when polished.

This photograph shows the interior of the West Rutland marble mill. The length of the shop was 1,000 feet. Blocks were brought into the mill by railcar and were then sawed into smaller slabs, often by using diamond saws after World War I. These saws had 125 diamonds in their rims, and quickly sawed through the hard stone by making 300 to 600 revolutions per minute. Diamonds revolutionized stone carving. As the diamonds wear down, so too does the metal surrounding them, which then exposes more embedded diamonds. Such blades could be as much as nine feet in diameter and were mounted side-by-side to create multiple cuts in one run.

Above, men work on rubbing beds in the finishing shop in Proctor. Their job was to smooth out the saw marks and irregularities as the huge slabs of marble came off the gang saws. The rubbing beds consisted of large, flat, round discs on which the slabs were placed. Sand and water were mixed together and trickled onto the disc, acting as an abrasive to smooth down the stone's surface. Once smooth, a slab was ready to be hand-polished to its final finish. The polishing mill at the Proctor shops is below.

After the marble was smoothed on the rubbing bed, if its final fate was to be a column or other circular piece, it was turned on a lathe and then planed to smoothness. In the exterior building shops in Proctor, marble workers cut fluted marble column with the planing machine. The column drums seen here were destined for the US Supreme Court building. Long smooth strokes of the planer rounded the corners and produced perfect flutes. When America engaged in World War II, the marble plant in Proctor transformed a large portion of its operations for war industrial purposes. The same planing machines that molded the massive marble sections for the Supreme Court, the National Gallery of Art, and the Jefferson Memorial were now enlisted in the cause of war. Airplane parts, ship winches, and weapons were manufactured alongside marble memorials in the huge monument shop.

Above, large marble columns are polished on lathes. Great care was taken during the polishing routine to ensure proper fitting with the joints when the columns were set. After polishing, the marble column was then set on yet another lathe so a carborundum saw could make a true cut for the joint, to ensure the proper fit. Marble columns were often used in public buildings in the early-to-mid-1900s to signify strength and trust in the businesses established within the stone edifice. Below, a man poses with a column produced for the post office in Camden, New Jersey.

In 1888, Alton J. Shaw developed the first electric overhead traveling crane, and established the Shaw Electric Crane Company in Muskego, Michigan. Shaw employed about 400 men and manufactured a large line of overhead cranes. They provided many types of cranes to the Vermont Marble Company over the years. Electric cranes were used in the marble yards as well as inside the shops to move heavy blocks of marble from place to place in the production process. Cranes were the true workhorses of the marble company mills and shops, moving blocks weighing many tons from the quarry to the marble yards and throughout the finishing processes. Finally, traveling cranes were used to load the finished marble products—columns, building stones, and memorials—carefully onto railcars, where they were transported to their final destinations throughout the country.

Redfield Proctor traveled to Carrara, Italy, to find artisans, trained through the generations to sculpt its white marble, who were worthy of carving Vermont's pure stone. After contacting the American consul in Italy, Proctor convinced five Italian sculptors to come and work for him in America. He arranged their passage overseas. Once established in their new life, the carvers repaid Proctor through deductions from their wages. These first Italian immigrants were well-trained experts and were recognized as superior sculptors, craftsmen, carvers, and stonecutters. As demand grew for increasingly ornate headstones and interior marble work, the sculptors took on apprentices and taught them the art of marble carving. Between 1882 and 1894, a steady procession of Northern Italian craftsmen followed the first immigrants, and the Rutland region soon became their home.

Much of the repetitive molded carved work, as seen, for example, on a building's cornice and moldings, and on other curved work, was done on the carborundum machine. If there was only a little cutting to be done, or if it was a complicated pattern, it was executed by hand with pneumatic tools. In this photograph, a carver using a pneumatic chisel makes fast progress shaping stone. Pneumatic carving tools work by placing many thousands of impacts per minute upon the end of the tool. This created the ability to "shave" the stone, providing a smooth and consistent stroke, and allowed larger surfaces to be worked. The carvers used hand tools such as lettering chisels to form letters in the stone, and fishtail chisels to make pockets and valleys in intricately carved sculpture. Letters were also cut onto marble by sandblasting. Metal letters were glued onto the stone, then a sand blast directed at high pressure on the surface quickly cut down the stone except where it was protected by the steel.

Starting with a large block of marble, the high points of the sculpture were roughed out down to the important fixed points. From there, the final details were carved. Carvers used three basic types of chisels. At left, a carver uses an air-powered point chisel to rough out the design in the stone. Toothed chisels would then be used to smooth the points down and begin to create curves in the marble. Finer chisels shape the finer details of the carving.

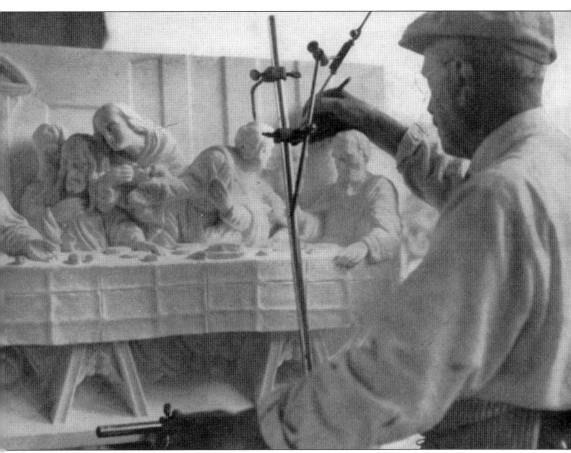

Italian sculptor Lando Bardi works on a section of a Last Supper panel using a pointing machine in 1938. Bardi was a native of Carrara, Italy, and a graduate of the Royal Academy of Art. He went to the Royal Academy at age 14 and remained there for five years. During that time, he studied modeling, carving, architecture, mathematics, and history. He came to America in 1893 and worked as a sculptor in Washington, DC. Redfield Proctor discovered him working as a carver in Washington and offered him a job at the Vermont Marble Company. Bardi moved to Vermont and made his home in Proctor, working at the company for 43 years. He was a mentor and a teacher of sculpting techniques for many of the carvers at the Vermont Marble Company.

Many thousands of blueprints for tombstones, buildings, and memorials were created in the drafting room. The Vermont Marble Company required certain specifications and detailed drawings to be submitted to their drafting department by the architects who designed the projects. These drawings had to include, in plan and elevation, the dimensions of all stones, positions of joints, and the spacing of all repeated ornamental work. Teams of drafters then worked together to create the plans that would eventually be made into gleaming pillars, buildings, cornices, and more at their final destination. In the shops, all cutting had to be done in accordance with the drafted drawings. Each piece of marble was planned to carefully fit with the next, so that on the building site, the marble would fit together seamlessly. The company even went as far as to specify that all marble stored at the building site had to be off the ground and protected from any staining before it was set into the building.

The Vermont Marble Company boxing room is seen above. In 1922 alone, more than six million square feet of lumber was used for boxing finished products for transport. The lumber was also used for the company's various building expansions, additions, and renovations. The lumber was harvested from the company's vast quarry land holdings throughout the United States. Below is a photograph of the Vermont Marble Company lumber camp at Pico Mountain, east of Rutland. The company owned acres of land on Pico Mountain, part of which was a camp owned by the Proctor family, who brought guests to enjoy hunting and other recreational activities.

In 1942, during World War II, the Vermont Marble Company transformed a large portion of its operations for wartime industrial manufacturing. The photograph above shows seam-welders and presses used to manufacture 155-millimeter shell cases. The photograph below shows one of the marble shops that was converted to wartime work. Even while wartime production was taking place, the marble company continued its work building tombstones and memorials. As men were called to war, women took their places at the machines.

During the war, the Vermont Marble Company produced various war machinery and components, including shell cases, furnace chargers, and the ship winches that were used to lower the landing craft from ships during the D-Day invasion on the beaches of Normandy. The exterior shops housed the planing machines and were converted to manufacture engine bases for Liberty ships and massive castings for machine tools. The marble workers learned to operate machine planers, milling machines, drills, and boring mills. Much of the marble machinery was converted to war work: carborundum moulders became milling machines and diamond saws were made into drills. Women were brought into the shops to run cranes and machine tools, work in the shop offices and stockrooms, and operate many of the woodworking machines. Many women worked in the mica room, turning rough mica into finished film for use in radio parts. For its wartime production efforts, the Vermont Marble Company was awarded the Army-Navy "E" Award three times during the course of World War II.

After World War II, business practices began to change at the Vermont Marble Company in response to the times. A modern era was ushered in. Second and third generations of the original immigrant workers continued to work the marble, but their roles became subtly different. More emphasis was made on finding new uses and markets for Vermont marble. Tombstones and memorials were still in demand, but demand for heavy blocks of building stone began to diminish. Thin interior tiles and marble novelty gifts became a staple of production. In this 1945 photograph, descendants of marble workers, now working for the company too, pose at the head gates of Sutherland Falls. From left to right, they are Rollin Champine, Dino Baccei, Mike Lewis, and Carl Steel.

Six

LIFE IN TOWN

The town of Proctor's marble bridge has been the centerpiece of the village since 1914 and is still in use today, connecting the east and west sides of town. Other notable landmarks in town are the white marble St. Dominic Catholic Church, the rock-faced marble Union Church, and the white marble Proctor Junior-Senior High School.

The town's early settlers traveled to Rutland or Pittsford to buy the goods necessary for their survival at the foot of Sutherland Falls. As the marble industry grew and houses rose on vacant lots, the shopkeepers of Pittsford aggressively sought the business of the townspeople of Sutherland Falls. They sent teams twice weekly to Proctor, taking orders on one trip and delivering the next. In the 1860s, two partners, Mr. Haywood and Mr. Hall, built a store right on the grounds of the marble yard. The partners owned the stock of the store but rented the building from the marble company. The store provided groceries, dry goods, drugs and medicines, animal feed, and other essentials. This 1870 photograph shows the village store to the left of the Sutherland Falls Marble Company mill, which at that time was still housed in a single building.

The village store burned down in 1872, losing all business records and details of its construction in the blaze. The Vermont Marble Company then took control of the store. On December 16, 1872, H.E. Spencer opened the store again in the newly rebuilt building. By the time the mill whistle blew signaling the end of the day, the doors were open and the store was parceling out provisions to its first customers.

The store stocked groceries and had a furniture department, a tin shop, and other provisions. By 1880, it housed the library and the Odd Fellow's hall on the second floor. Garron's Barber Shop was in the basement along with the post office until the completion of the Post Office Block in 1910.

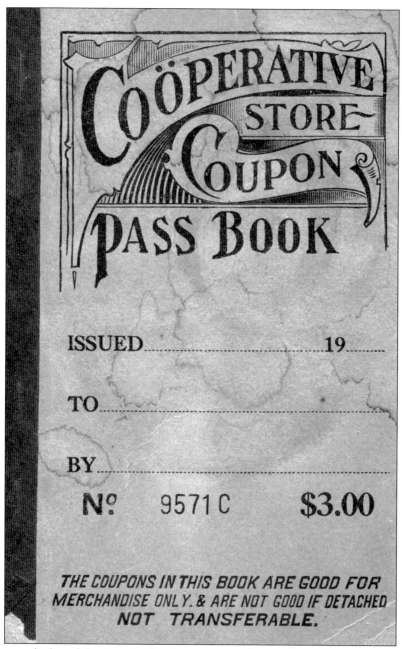

COÖPERATIVE STORE COUPON PASS BOOK

ISSUED.. 19..........

TO ....

BY ....

Nº   9571 C $3.00

THE COUPONS IN THIS BOOK ARE GOOD FOR
MERCHANDISE ONLY. & ARE NOT GOOD IF DETACHED
NOT TRANSFERABLE.

There was very little cash business in the company-owned store. Company "scrip," or passbooks, were used for credit in the store. At the end of the month, the amount owed was deducted from the men's wages, and sometimes in lieu of pay, men received a red pay envelope, indicating that a balance was still due to the store. In 1903, the store became a company cooperative and the Vermont Marble Company did not earn profits from the store. All profits were divided at the end of the year among the store's patrons. The company felt that by doing this, it eliminated the public perception that it had a selfish interest in having employees trade at the store. In the first year of this arrangement, the employees trading at the Proctor store received a dividend of 10 percent, which was paid to them in January 1904 by bank check.

The Post Office Block was another center of commerce. Seen here in 1910, the building also housed a drugstore on the lower left and a barbershop on the lower right. The second story was home to a clothing store on the upper left and the post office on the upper right. On the third floor, there was an insurance company, a millinery, and dental offices. The top floor was the Masonic IOOF hall. The town of Proctor was a vibrant community of about 2,500 that supported the community businesses. Almost all the townspeople's needs could be met right there in the center of Proctor.

The store in the middle of the marble yard caught fire on November 11, 1913. As it burned to the ground, the goods that could be saved were moved to locations throughout the town. Until the new store was built in 1915, the grocery department was housed in the old library building, the drugs and medicine were housed at the Post Office Block, and dry goods were sold from the village hall basement. Every effort was made so that townspeople could meet their daily needs despite the lack of a centrally located store.

The new store, built in 1915, was one of the best-equipped and most modern buildings of its kind in New England. Built of tapestry brick and trimmed with white marble, it even had an underground refrigerating plant. The store was especially proud of the Brunswick Cooling System and advertised that green vegetables were kept fresh. The cleanest of clean groceries were displayed in the refrigerated section. The fixtures in the store were elegant and modern, purchased from the Grand Rapids Show Case Company. The interior wall finish was of a marble white color, with a trim of hard pine stained to match the fixtures. The new store was still run as a cooperative, and from the earnings of the year, all operating expenses—rent, merchandise, and salaries—and interest amounting to four percent of the capital were deducted. The remainder was divided among the employees, with the share amount dependent on each individual's aggregate purchases in the year.

The Vermont Marble Company named a committee of five employees to confer with company management on matters of policy, the distribution of profits, and the types of goods sold. These photographs of the company store in 1930 show the wide variety of merchandise available to its patrons. Prices of goods were as low as those in the surrounding towns and in the city of Rutland. At its grand opening on Saturday, March 6, 1915, the store advertised fresh produce in its grocery department, men's furnishings and shoes, hardware, drugs and medicines, crockery, wallpaper, and dry goods. The selection was so complete that townspeople had no need to travel to Rutland to purchase goods.

Proctor Park is seen above in 1881. The original stone schoolhouse in the background was built in 1866 and used until 1903. The Union Chapel, on the knoll, was built in 1880 and burned in 1889. The present Union Church was built on the same site in 1891. The photograph below shows the same area 30 years later. The Post Office Block (left) and the town hall (center) were both built in 1910. The YMCA building (right) was erected in 1903 and became the Sutherland Club in 1919.

This suspension footbridge was built in the late 1880s over Otter Creek, above Sutherland Falls. Swaying in the wind, and open in all seasons of the year, the swinging bridge provided workmen living in the northeast part of town—the Patch Street and Williams Street areas—with a shorter route to their jobs at the marble company and the quarries. Otherwise, they would have to walk down through the covered bridge, back along the railroad tracks, and over to the company buildings. The sign reads, "This bridge is intended solely for a footbridge and all running, scuffling, or climbing on it is absolutely forbidden." In the 1930s, the bridge was renovated with new cement posts and a new floor. The bridge was supported only by seven guy wires, four of which also served as hand railings.

This 1897 photograph shows a man on the footbridge over Sutherland Falls. The view of the falls and the mountains and valley to the north was spectacular from the center of the bridge. The bridge led to a grove on the east side of the falls, which, it was thought, would become a place for picnics and outings. A man named Peter Morganson is credited with supervising the building of the bridge. In 1912, the suspension bridge was condemned and a new and stronger one was built to replace the old one. Another suspension bridge was built below the falls, joining the factory to the bank on the east side of the river. This bridge was not used for foot traffic, but instead to transport marble slurry waste, a mixture of sand, water, and marble left over from production processes, over Otter Creek, where it was deposited in a nearby swamp.

Proctor's east and west sides were connected by a covered bridge spanning Otter Creek. Three different covered bridges were built in this spot over the years. The first was built in 1794 and replaced in 1811. The final covered bridge was built in 1839. The bridge was supported by town lattice trusses and was distinguished by unusual arched portals and long windows. It was replaced in 1914 by a marble-arched bridge.

This photograph from around 1905 looks west at the covered bridge and the Canoe Club on Otter Creek, as men and women enjoy a canoe outing on the creek. The Canoe Club was built in 1905 to the south of Sutherland Falls on the east side of Otter Creek. Originally, it held 10 canoes, but it was so popular that the next year the building was expanded to hold 20 canoes and four boats. The Canoe Club charged an initiation fee of $2.50 and monthly dues of 50¢.

In 1914, the covered bridge spanning Otter Creek was torn down, and a magnificent marble bridge, the centerpiece of Proctor, was built. The bridge was commissioned by Redfield Proctor's wife, Emily, in memory of her son, Gov. Fletcher D. Proctor. Designed by architect Harry Leslie Walker of New York, the bridge was built of concrete, with marble facings and trimmings. Intricate details of this bridge include the marble pedestrian railing, made of 288 turned marble posts. The marble bridge was renovated in 2002.

The Vermont Marble Company enjoyed a reputation for innovative industrial healthcare. The first hospital (above), east of Otter Creek, was built from a private home. It opened on August 6, 1896, and Ada Stewart was its first matron. It accommodated 10 patients. The operating room was finished in Vermont marble and supplied with every obtainable convenience. Below, a group of nursing students are on the lawn in front of the hospital. Along with the matron, the hospital staff consisted of two nurses, two student nurses, two attending physicians, and four consulting physicians and surgeons. Injured and ill Vermont Marble Company employees were treated free of charge. Other residents of town were also allowed to use the hospital's services, at a charge of no more than $4 per week.

The Vermont Marble Company, aware of the many dangers of the quarrying industry, was the first company in the nation to hire an industrial nurse. The nurse, Ada Stewart, began her work in 1895. Before the hospital was built, Stewart cared for patients in their homes. She often used a bicycle to make her way around town to visit patients, which was a great novelty for the recent immigrants, and she spoke to them in a combination of sign language and a smattering of their native tongues. In addition to caring for injured marble company workers, she cared for their families, helping in emergencies, dressing wounds, and teaching ways of health and good habits. She gave a weekly health talk at the Proctor school. When the hospital was built, it also became home to a nursing school. This photograph shows Proctor Hospital School of Nursing graduates Mable Stevens (left) and Mary Candon in 1904.

The original hospital was soon outgrown, and a new one was built in 1904 on Ormsbee Avenue. It was built of rock-faced marble and was well equipped and splendidly staffed. This hospital was large enough to serve not only the residents of Proctor but also the surrounding community.

This photograph of Proctor Hospital was taken in 1965, three years before it closed. The hospital was filled with interior marble finishes and demonstrated how marble's sanitary properties were good for hospitals. The Vermont Marble Company marketed the light-diffusing properties of its polished white marble as perfect for operating rooms. The nation's architects used Proctor Hospital as a prototype for hospitals across the country, and marble was used extensively in hospital interiors in the early 1900s.

The Vermont Marble Company founded a chapter of the YMCA in Proctor in 1903. The company built the rock-faced marble building (above) at a cost of $36,000 to house the YMCA, and invited Christian men of all nationalities to join. Many classes, sporting teams, and activities were offered to educate immigrants in the ways of America. Below is the 1908–1909 YMCA basketball team, from left to right, "Punch" Daley, William Donnelly, Marcus Ling, Al Barr (holding basketball), Bernie Johnson, Robert Hart, and Peter Erickson.

Although Catholic men were allowed to join the Proctor YMCA, Bishop Joseph Rice of the Catholic diocese frowned upon Catholics joining a club with a religious affiliation outside Catholicism. Father Crosby of St. Dominic Church approached Vermont Marble Company officials about changing the club's affiliation from the YMCA in 1919. The company agreed, as it wanted employees of all faiths to freely gather together socially. The club's membership became open to every man in town. The club was changed from a YMCA to the Sutherland Club, a sports and social club with no religious affiliation.

The women of town had their own social center. Emily Proctor, like many young women of her time, was well educated and looking for a way to be active in public life. She decided to address the needs of the immigrant women of her town and started Cavendish House, based on the settlement houses popular with progressive reform, in January 1910. There, immigrants were taught to survive and prosper in their new world. Classes were taught in cooking, sewing, English, surgical dressings (during World War I), physics, household accounting, history and geography, hygiene, millinery, basketry, typewriting (in the summer of 1914), and physical culture. Bath and shampoo rooms were offered free of charge to all women and children who wished to use them. Cavendish House had a governing board of one man and two women. The staff consisted of six women: a physical education director at the public school, a school nurse, a nurse for classes and miscellaneous work, and three domestic science teachers.

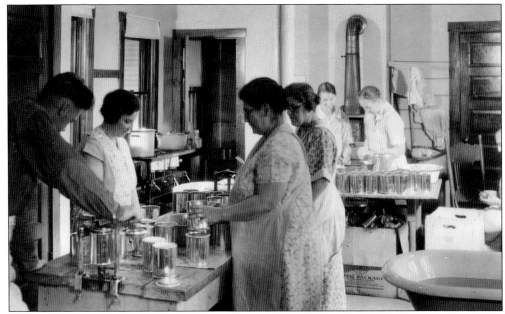

The work at Cavendish varied from year to year to meet the needs of the times. Gardening was taught, summer school was held for pupils who did not pass in the public school, and cooking was done to help out sick families. Seen in this 1935 photograph of a "Welfare Canning Project" are, from left to right, Leah Franzoni, Mrs. Mitchell Frappier, Priscilla Wood Parker, Mrs. Joseph Cizmedia, and Mrs. George Bisczco.

Canned goods preserved at Cavendish House are loaded for delivery to needy families throughout the area in this 1934 photograph. Emily Dutton Proctor exemplified a long tradition of upper-class elites who had a strong sense of social duty to the poor.

The Proctor public library (above) was built in 1891 out of stones from the old schoolhouse. The building served as a library for more than 20 years. The new Proctor Free Library (below) opened in May 1913 on the banks of Otter Creek. Its red brick exterior trimmed with white marble was built as a gift to the town by Emily Proctor in memory of Arabella Proctor Holden. The former library had its interior remodeled and was then presented to the village on May 20, 1920, as a municipal memorial and home to the American Legion. The building currently houses the Proctor town offices.

The stone schoolhouse at left was built in 1866. Masons from Rutland laid the stone and J.L. Patch and Sons created the woodwork. The keystone over the door, which reads "A.D. 1866," was carved by B.F. Taylor. The schoolhouse was built on marshy ground in the town's center. In 1919, a new high school (below) was built, which serves as Proctor Elementary School today. The cost of the building was $60,000, and the work of grading and equipping the grounds on the top of the hill was a private contribution to the Proctor school system from Emily D. Proctor.

Seven

ACROSS THE NATION

The Tomb of the Unknown Soldier was carved in West Rutland and Proctor from a block of Yule marble. It was crated and placed on a railcar for shipping from the Proctor shops of the Vermont Marble Company. The sign on the side of the crate was a marketing strategy for the company. As the train rolled through towns and cities on its way to Washington, DC, it spread the message that Vermont marble was the preferred choice for national memorials, and should be each person's choice for their family memorials as well.

Marble from Vermont graces many buildings and monuments in Washington, DC, including the US Senate office building, seen here, which was completed in 1908. Sen. Redfield Proctor's active political career resulted in many building projects throughout Washington, DC, featuring Vermont marble.

Memorial Continental Hall, the oldest building at the Daughters of the American Revolution complex, was designed in 1905 by prominent Washington architect Edward Pearce Casey, the designer of the interior of the Library of Congress. Upon its completion, Memorial Continental Hall was quickly recognized as one of Washington's most elegant buildings.

Memorial Amphitheatre, Arlington, Va.

The Arlington Memorial Amphitheater was built in 1915. A beautiful memorial to the deceased heroes of our nation, it is situated in Arlington National Cemetery in Virginia and faces the Lincoln Memorial across the Potomac River in Washington, DC. In all, 87,000 cubic feet of Mountain White Danby marble was quarried, finished, and shipped for use on the amphitheater. The cost of construction was $750,000. The building seats 5,000 people in the amphitheater and several thousand more in the colonnade.

The American Red Cross Building in Washington, DC, was completed in 1917. This memorial building was built to honor all the women of the Civil War, Union and Confederate, who cared for the men who were wounded during the war. Congress set aside $400,000 for a building and a site, but required matching funds of at least $300,000. Recognizing the importance of this memorial building, and in support of Red Cross efforts, private donors stepped up with contributions amounting to $400,000. More than 200 carloads of Vermont marble were used in the project. The exterior's elegant facade is of Danby white marble. The interior entrance corridor, highlighted by a marble staircase, is built of White Cloud Rutland marble and accented by columns of Royal Antique Green.

In 1932, construction began on the US Supreme Court building. The Vermont Marble Company was tasked with creating the 16 entrance columns. Each is about 52 feet high and constructed of drums, with a diameter at the base of six feet. To provide 80 of these huge fluted drums was a task of real magnitude and source of pride for the company's workers. The scene in the marble company shops in those days seemed to be one of disorder. Columns, bases, and cornices were spread about as they were worked on by teams of men. Once the pieces reached their final destination in Washington, they were pieced together and set into the building site.

Drums are matched and fitted together in three pieces as columns for the Thomas Jefferson Memorial in the Proctor marble yard. These columns of Vermont Danby marble make up the colonnade around the Jefferson Memorial. A total of 353 carloads of high-grade white marble were used in the project. More than one-third of this quantity went into the great columns that surround the structure.

A colonnade of Ionic columns made of Vermont marble circles the Jefferson Memorial in Washington, DC. Seen in the marble yard of Vermont, each drum piece for the Jefferson Memorial columns was slightly taller than a man. The shafts of these columns consisted of 324 drums approximately five feet in diameter and six feet high. Each of these pieces represents a fair-sized quarry block.

The Lincoln Memorial was built of marble from the Yule quarry in Colorado, which was owned by the Vermont Marble Company. The Lincoln Memorial was completed in 1922. Each one of the 36 Doric columns represents one of the 36 states in existence when Lincoln was president. The columns support an entablature with a frieze of 48 sculptured white marble festoons separated by eagles, representing each of the states in existence in 1922.

The Curtis Building rose on the west side of Independence Square in Philadelphia, echoing Independence Hall in its choice of materials: red brick and white marble. Curtis Publishing's flagship publications were *Ladies Home Journal* and the *Saturday Evening Post*. Edgar V. Seeler was the architect and Frank C. Robert & Co. were the engineers. It is divided into four distinct structures: the publication building; what was called the convenience belt, behind the publication building; the manufacturing building, in two sections; and the power building. The interior of the block had a center court to provide light and air.

The Curtis Building was a significant project for the Vermont Marble Company, and its progress was well documented. The cornerstone was laid in 1911, and the building was completed in successive stages over a 10-year span, ultimately completed in 1921. The publication building was entered through bronze gates set back behind 14 monolithic columns of Vermont marble. It housed the bookkeeping, circulation, administrative, and advertising departments, with the editorial offices of the *Journal* and the *Post* on the sixth and seventh floors. The upper floors contained the women's lunchroom, a library, and a hospital.

Above, the monolithic columns for the Curtis Building are loaded onto railcars to be shipped to Philadelphia. The photograph below shows the columns for the Curtis Building in several stages. One of the rough shafts is in the grip of a traveling crane, and some of the finished columns have already been loaded onto the railcars. The columns were cut from the Vermont Marble Company's Hollister quarry, a few miles from Proctor in Pittsford. The columns weighed 50 tons each, were 29 feet, 9 inches long, and about three feet in diameter.

On March 4, 1921, Congress approved a resolution providing for the burial of an unknown and unidentified American soldier from World War I in the Arlington National Cemetery Memorial Amphitheater. The simple white marble tomb, placed over the resting place of the unknown soldier, was fabricated from Vermont Danby marble and was later used as the base for the carved, larger tomb that was completed by 1931. The block of pure white Yule marble for the tomb was fabricated into the famed sarcophagus, with some carving done on-site at the West Rutland and Proctor workshops of the Vermont Marble Company. The finish details were carved after the tomb was in place at Arlington National Cemetery.

Above, just before crating for shipment, local members of the American Legion pose with the Tomb of the Unknown Soldier in honor of its significance to the veterans of our nation. The photograph below shows the finished Tomb of the Unknown Soldier at the Arlington Memorial Amphitheatre in 1934. As people came to pay their respects, the need grew for a formal guard program. In 1937, the tomb was placed under guard 24 hours a day, 365 days a year, but it was not until 1948 that the 3rd US Infantry Regiment BCT, "The Old Guard," began guarding the tomb in concurrence with their mission of conducting ceremonies to maintain the traditions of the Army, showcase the Army to our nation's citizens and the world, and defend the dignity and honor of our fallen comrades.

Vermont marble sculptors worked on monuments, memorials, building details, fine interior statues, altars, and friezes that went to cities across the nation. The most famous buildings in Washington, DC, owe their beauty to the carvers from Carrara, their descendants, and other marble company workers who were fortunate enough to learn the trade from the Italian master carvers. Here, sculptor Renzo Palmerini works on the Majesty of Law sculpture in 1964. Palmerini was born in 1940 in Pietrasanta, Italy. He graduated from the Institute of Art in Pietrasanta and came to America as a young man. He worked at the Vermont Marble Company for more than 25 years as a marble sculptor and a draftsperson. He was one of the last of the Italian sculptor carvers to work at the Vermont Marble Company. Palmerini is retired and still lives in the town of Proctor today.

Renzo Palmerini carved an 80-ton block of marble into the statue named Majesty of Law, which sits at the right entrance of the US House of Representatives office building. The figure holds a book of federal laws with the US seal on the cover and a sword representing valor. It faces the Spirit of Justice statue, which was also worked on by Palmerini. Both the Majesty of Law and the Spirit of Justice sculptures are credited as the work of artist C. Paul Jennewein. Palmerini and Francesco Tonelli, like other master carvers of the Vermont Marble Company, were trained production carvers who took an architect's or designer's models and specification plans and transferred them into stone to a tolerance of 1/32nd of an inch.

In 1962, Renzo Palmerini and Francesco Tonelli created the Spirit of Justice sculpture, which sits on the left side of the entrance to the House office building in Washington, DC. The cornerstone of the building was laid in May 1962, and it reached full occupancy in February 1965. Its total area is 2,375,000 square feet. The figure holds a lamp representing truth and righteousness. Her hand rests gently on the shoulder of a small boy, symbolizing justice moderated by love. Both statues were completed by 1964. Palmerini, Tonelli, and others in the group of old-time carvers from Italy worked on many of the additions and renovations to buildings that were originally constructed out of Vermont marble, including the Capitol building, the Senate office building, and the Library of Congress.

As building styles changed after World War II, the Vermont Marble Company began to offer thin marble-veneer slabs, which were often combined with glass and steel for exterior buildings. One of the first postwar skyscrapers, the Secretariat building of the United Nations complex, was completed in 1950. With its blue glass sides and marble ends, it was the most striking landmark on New York's East River at the time. The beautifully gleaming variegated Vermont Pearl marble was quarried in West Rutland. The proposed site for the construction of the building was purchased by John D. Rockefeller Jr. for $8.5 million and donated to the United Nations in December 1945. Construction began in 1948. One of the last great projects of the Vermont Marble Company, its completion signaled the end of an era. The company was purchased in 1976 and operations ceased shortly afterwards.

DISCOVER THOUSANDS OF LOCAL HISTORY BOOKS
FEATURING MILLIONS OF VINTAGE IMAGES

Arcadia Publishing, the leading local history publisher in the United States, is committed to making history accessible and meaningful through publishing books that celebrate and preserve the heritage of America's people and places.

Find more books like this at
www.arcadiapublishing.com

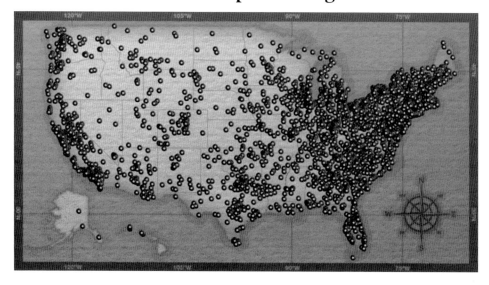

Search for your hometown history, your old stomping grounds, and even your favorite sports team.

Consistent with our mission to preserve history on a local level, this book was printed in South Carolina on American-made paper and manufactured entirely in the United States. Products carrying the accredited Forest Stewardship Council (FSC) label are printed on 100 percent FSC-certified paper.

MADE IN THE USA